COASTLINES
OF NEW ZEALAND
PHOTOGRAPHY BY ROB SUISTED
TEXT BY ALISON DENCH

CONTENTS

INTRODUCTION	4
NORTHLAND	6
AUCKLAND REGION	12
COROMANDEL & BAY OF PLENTY	18
WAIKATO & KING COUNTRY	24
TARANAKI & MANAWATU	26
EAST CAPE & HAWKE'S BAY	32
WELLINGTON REGION	38
NELSON & MARLBOROUGH	44
KAIKOURA	50
WEST COAST	52
CANTERBURY	58
OTAGO & THE CATLINS	64
SOUTHLAND	72
STEWART ISLAND/RAKIURA	76

INTRODUCTION

There are two distinctly different faces to the New Zealand coast, one looking to the Pacific Ocean and the other to the Tasman Sea. The deeply indented shoreline and sheltered sandy coves and beaches of the east are a playground and place of relaxation. But the west coast's long, straight, black-sand beaches, bleak in winter and scorching in summer, offer no haven from the prevailing winds and they are places to reflect rather than relax.

New Zealanders have always favoured seaside communities. Maori and European settlers looking for a ready food source found it in fish, seals, shellfish and crustaceans. Harbours, bays and river mouths provided ports for coastal shipping, the only easy method of transport before an inland road network was built. The fertile coastal plains could be cultivated for first kumara gardens and then farms.

The marine harvest quickly became exploitation. Maori hunted seals to near extinction, sealers and whalers slaughtered marine mammals for profit, and modern commercial fishing techniques threaten some species. Pounamu collectors and gold diggers pillaged the west coast beaches, and steel makers have mined the ironsands too.

But the coast is resilient and damage can be repaired. Regardless of human impacts, with each tide, each storm, each earthquake the coast changes. Sand is swept in and swept away, rivers carry rock debris down from the mountains, cliffs erode into arches and stacks and crumble into the sea, land is thrust upwards out of the sea by violent seismic forces. Climate change may be threatening to alter the coast forever, but that isn't new: rising sea levels after the last ice age drowned coastal valleys and created the Fiordland and Marlborough sounds and deposited golden sand on the beaches of the east coast.

Perhaps it is that very mutability that makes the New Zealand coastline endlessly unpredictable and endlessly rewarding to visit.

Left: A pohutukawa tree shades a family campsite at Fletcher Bay, in the far north of the Coromandel Peninsula.

NORTHLAND

Over the centuries Northland, with its subtropical climate, peaceful east coast havens and rich marine life, has been a magnet for human settlement of all kinds. Pre-European Maori found plenty to sustain them in the sea and in the forests, while in the Bay of Islands early European settlers found safety for their ships in the natural harbours, fertile soils for their crops and, to begin with at least, a warm welcome from Maori keen to trade with them. It became the first area of permananent settlement and the site of the first capital. These days it is holidaymakers that feel the pull from the north, drawn to the beaches and dunes, sheltered bays, historic harbours and the relative simplicity of life in the rural far north.

Left: Cape Maria van Diemen and, off the end of the point, the Motuopao Island Nature Reserve can be reached on a walkway from Cape Reinga (Te Rerengawairua).

Above: Kayakers explore the indented rocky shore at the southern end of Maitai Bay.

Right: The sheltered waters at Taipa, in Doubtless Bay, are perfect for children learning to sail.

Left: Spirits Bay (Kapowairua) is sacred to northern Maori as the place where souls gather before departing from Te Rerengawairua (the leaping place of the spirits).

Below: A memorial to the *Rainbow Warrior*, the Greenpeace protest ship bombed in 1985 by French agents, overlooks the ship's final resting place in Matauri Bay.

Right, clockwise from top left: The Hole in the Rock at Motu Kokako (Piercy Island), off Cape Brett; bottlenose dolphins ride a bow wave in the Bay of Islands; tiny fairy terns/tara-iti nest on the beach at Waipu.

Left: The villages of Omapere and Opononi occupy a thin ribbon of flat land beside the Hokianga Harbour. The harbour's full name Hokianga-nui-a-Kupe acknowledges the legendary Polynesian explorer Kupe, who left from there to return to his homeland.

Above: Pohutukawa tree at Rawene, Hokianga Harbour.

Right: The Hokianga Harbour's wharves, many of them now decaying, once served flourishing kauri, gum and flax trades.

AUCKLAND REGION

The coastline of the Auckland region is a study in contrasts. In the west straight-lining blacksand beaches, face to the weather and back to the rugged forests of the Waitakere Ranges, can feel bracing rather than welcoming, while in the east the bays, headlands, islands and golden sand beaches have an easy appeal to swimmers, boaties and fishers. The ever-growing metropolis of Auckland has its downtown on a narrow isthmus between two harbours, the small, sheltered, suburb-lined Waitemata to the north and the vast, wild, tidal Manukau to the south. Surrounded by water, Aucklanders certainly feel a close connection with the sea, enough to warrant giving their home the name of 'City of Sails'.

Left: Rangitoto Island, an active volcano across from the sought-after Auckland beach suburb of Takapuna, last erupted just 600 years ago.

Above: Yachts moored in Oneroa Bay, Waiheke Island. The islands and bays of the Hauraki Gulf are easily explored by boat and Auckland has the highest rate of boat ownership of any world city.

Right: Rangitoto Island has no permanent residents, but it is a popular ferry day trip from Auckland.

Above: Completed in 1959, the Auckland Harbour Bridge crosses the Waitemata Harbour, linking the central city with the north shore suburbs.

Left: Sir George Grey, New Zealand's governor in the middle of the 19th century, built a mansion at Kawau Island in the Hauraki Gulf and liberated wallabies, zebras and monkeys in the grounds. The zebras and monkeys are long gone, but the wallabies remain.

Right: Auckland, New Zealand's largest conurbation, is a harbour city with a sea port right at its heart. The lights of the CBD, the waterfront and the port illuminate the calm waters of the Waitemata.

Left: Lion Rock dominates the wild, wide sweep of Piha Beach. On the west coast between the Waitakere Ranges and the Tasman Sea, Piha is a popular spot for swimming in summer and walking all year round.

Right: A south-westerly prevailing wind is perfect for hang-gliding at Muriwai Beach.

Below: Remote, bleak Whatipu, at the mouth of the Manukau Harbour, has a broad beach of black ironsand backed by huge dunes and swampy wetlands. In 1863 New Zealand's worst maritime disaster occurred near here, when HMS *Orpheus* was wrecked and 189 lives were lost.

COROMANDEL & BAY OF PLENTY

Over the peaceful Coromandel, down the sun-soaked Bay of Plenty towards East Cape, the Pacific Coast Highway leads holidaymakers away from the bustle of Auckland to the ease and relaxation of the beach life. The Coromandel Peninsula, within sight of Auckland but a million miles away in spirit, has long been a weekend retreat for city-dwellers heading to campgrounds and basic holiday houses, but the soft sand beaches are alluring and the once quiet settlements are growing and the secluded beaches may not be so secluded any more. Further south, the citizens of Tauranga, living so close to the Bay of Plenty's long open stretches of spectacular ocean beach can, if they choose, live the beach life 24/7.

Left: At Cooks Beach, holiday houses sit cheek-by-jowl behind two kilometres of tempting white sand.

Above: The sparkling waters and soft pink-tinged sand of Cathedral Cove are protected by Te Whanganui-A-Hei Marine Reserve. Cathedral Cove is one of the Coromandel's most visited attractions, despite being accessible only on foot or by boat.

Right: The Coromandel coast is lined with evergreen pohutukawa trees, a species unique to New Zealand.

Above: At Hot Water Beach in the Coromandel thermal water bubbles up through the sand and holidaymakers dig their own natural hot tubs.

Left: Whiritoa, with its roving sandbars, is good for surfing.

Right: The Whangamata Harbour is a haven for yachts and gamefishing boats. The waters off the Coromandel peninsula are home to record-breaking marlin, spearfish, tuna and sharks.

Left: A sandy ocean beach runs unbroken for 25 kilometres south-east from Mount Maunganui.

Above: At the end of a narrow sandbar, the volcanic cone of Mauao offers a viewpoint over the town of Mount Maunganui. The busy summer resort town has both an ocean beach known for excellent surfing – and dangerous rips – and a sheltered harbour beach.

Right: A plume of sulphurous steam rises from Whakaari/White Island. Rising out of the Bay of Plenty 50 kilometres from Whakatane, Whakaari is part of the Taupo Volcanic Zone and is New Zealand's most consistently active cone volcano.

WAIKATO & KING COUNTRY

The coast of Waikato is largely inaccessible and thus largely undeveloped. For 50 kilometres from Waikato Heads, where New Zealand's longest river empties into the sea, to Raglan Harbour (Whaingaroa) no public road reaches the desolate coast and the only way to visit is by driving along the blacksand beach itself. At Raglan the coastline makes a break inland in a series of harbours ringed by settlements. Most are small and sleepy but Raglan is a fair-sized fishing town and seaside resort with quirky cafés that, happily for city workers, is also within commuting distance of Hamilton.

Left: The left-hand point break at Manu Bay, near the relaxed seaside town of Raglan, started attracting surfers from all over the world after it featured in the movie The Endless Summer.

Above: The ironsand beach at Marokopa is too wild for swimming, but perfect for surf casting for fish.

Right: This recreational fisher at Marokopa has made a good catch of kahawai.

TARANAKI & MANAWATU

Mount Taranaki, the dormant volcano that is the single dominating feature of the western coastline of the central North Island, soars above the region on its own peninsula and dominates the vista from every direction. There are a number of surf beaches around the peninsula's coast within easy distance of New Plymouth, a city once dismissed by some as a backwater but now coming into its own. Further south the wide arc of the South Taranaki Bight, which bears the brunt of the Roaring Forties winds, has no natural harbours – just kilometre upon kilometre of black volcanic sand backed by marine terraces – forcing the coastal shipping of the past to make do with river ports.

Left: Between Tongaporutu and Pukearuhe a long line of bluffs forms a rapidly eroding bastion against the Tasman Sea. These fossil sea-cliffs have been cut by wave action from marine terraces that were raised from the sea 120,000 years ago.

Above: The silhouette of Mount Taranaki (Mount Egmont) towering over the coastal plains is a reminder of the North Island's volcanic past and present. In Maori mythology, Taranaki lived in the middle of the island until being forced to flee after a titanic struggle with Tongariro.

Left: New Plymouth's popular waterfront walkway stretches for over 12 kilometres from Pioneer Park to Bell Block beach. Along the way it meanders past Len Lye's red Wind Wand, a 45-metre kinetic sculpture that bends and sways dramatically in the breeze.

Above: The tiny informal settlement of Tongaporutu, in northern Taranaki, nestles beside the river from which it takes its name. This relatively unmodified river mouth and nearby wetlands provide a home for a large number of coastal bird and fish species.

Above: The Awakino River, noted for its back-country trout and whitebait fishing, runs in to the Tasman Sea beside the baches of Awakino. In the early 20th century, when inland road travel was difficult, Awakino served as a port.

Right: At Waikawau heavy black ironsand lies below a buttress of orange sandstone.

Far right: In 1942 a German navy shipping mine broke from its mooring off Australia and washed ashore at Mokau. It was put on display – after 204 kilograms of TNT had been removed.

EAST CAPE & HAWKE'S BAY

The shoreline of the East Cape down to Hawke's Bay features heavily in the early history of New Zealand. It was here that in 1769 James Cook became the first European to set foot in Aotearoa, here that the first and fatal contact with Maori was made, here the disenchanted navigator found 'no one thing we wanted'. Maori resisted the intrusion of Pakeha for generations, and it was a century before the city of Gisborne could be founded. Today the East Coast, with its stunning beaches and a scattered population that retains a firm grip on its Maori heritage, still feels isolated, while neighbouring Hawke's Bay, source of some of the best wine and food in the country, is more urban.

Left: Pohutukawa trees are a feature of the East Coast. This coastal species flowers in December, giving it the nickname of the New Zealand Christmas tree.

Above left: Mount Hikurangi (1752 m), the highest peak in the Raukumara Range, looms over Tolaga Bay.

Above right: A small community has grown up at remote Waihau Bay, on the East Coast, since the land was subdivided in the 1990s.

Far left: A historic wharf, much loved by recreational fishers, stretches 660 metres into Tolaga Bay.

Left: Horses are a popular way to get around the mostly Maori settlement of Tokomaru Bay. The village is home to a number of artists and artisans, and is known as the craft centre of the East Coast.

Above: A statue in Gisborne honours navigator James Cook. The naval captain became the first European to make landfall in Aotearoa when he came ashore directly across the Turanganui River mouth on Kaiti Beach.

Left above: During the summer school holidays families descend on Anaura Bay for long days of camping, swimming, walking and just lazing in the sun.

Left below: The Waipatiki estuary – the name means 'water of the flounder' – was an important food source for Maori until 1931, when the Hawke's Bay earthquake raised the land and river flats and a surf beach was created.

Right, clockwise from top: Cape Kidnappers was so named by James Cook after local Maori tried to kidnap a crew member of his ship *Endeavour* in 1769; David Trubridge's Millennium Sculpture on Napier's Marine Parade indicates the place where the sun broke the horizon at dawn on 1st January 2000; there has been a colony of gannets/takapu at Cape Kidnappers since the 1870s.

WELLINGTON REGION

When the New Zealand Company selected the site for Wellington, the settlement agency noted the excellent deep-water harbour and turned a blind eye to the contours of the land. As a result, Wellington is crammed precariously into the hills around the harbour, sea views are dime-a-dozen and residents enjoy a good selection of city beaches. A world away – in spirit if not in distance – is neighbouring Wairarapa, where a string of tiny villages dot a savage coastline and where the few commercial fishers must risk the destructive Cook Strait storms without a sheltered port.

Left: Shipwrecks were commonplace on the exposed southern shore of Wairarapa before the lighthouse at Cape Palliser was built in 1897. The light of the beacon slowed but could not stop the run of wrecks in the most treacherous bay in New Zealand.

Above: At Castlepoint a jagged reef and spectacular rocky headland protect a sandy lagoon, creating a safe swimming beach on the windswept east coast. The holiday settlement has held annual horse races on the beach for over a century.

Left: The harsh conditions and rugged volcanic landforms of the Wairarapa coast did not deter early Maori, who fished and grew kumara on the eastern side of the Palliser Bay in the late 14th century.

Above: The Rimutaka Cycle Trail, a 115-kilometre route that largely circumnavigates the Rimutaka Range, crosses Mukamukaiti Stream near Windy Point on the shore of Palliser Bay. Here the track follows a shoreline that was raised by 2.7 metres in the 1855 Wairarapa earthquake, the most severe earthquake in New Zealand since European settlement began.

Right: Whitireia Park occupies the hilly headland that shelters Porirua Harbour from the Tasman swells. This easily fortified rock-fringed peninsula was occupied in the 19th century by Te Rauparaha and his Ngati Toa tribe, who migrated from Kawhia and conquered much of the Wellington region.

Above, clockwise from top: Wellington has of necessity spread from the limited flat land around the harbour to the surrounding hills; young sailors train in dinghies at sheltered Evans Bay, Wellington Harbour; on Wellington city's waterfront swimmers make the most of a still summer's day while the Max Pattie sculpture Solace in the Wind, a man serenely facing into a gale, reminds strollers of Wellington's reputation as the windy city.

Right: The inner suburbs of New Zealand's compact capital city, Wellington, are hemmed in between Lambton Harbour and its outer green belt of open hills, undeveloped save for a line of wind turbines. Much of the waterfront has been built on land reclaimed from the sea in a bid to increase the amount of usable flat land.

NELSON & MARLBOROUGH

The most northerly regions of the South Island are blessed with golden sand beaches in a multitude of bays and inlets, clear turquoise waters and mild weather. With four marine reserves, three national parks and two Great Walks, Marlborough and Nelson are a haven for anyone trying to get away from it all in the heart of New Zealand. The coasts of the two former provinces are quite distinct and offer different things to the holidaymaker. Marlborough's hills separated by drowned river valleys are perfect for sailing and kayaking, and Nelson's two huge, shallow bays protect sandy, safe beaches made for families.

Left: Farewell Spit, New Zealand's longest sandspit, curves away from the tip of the South Island to form the northern side of Golden Bay. The spit's wildlife reserve protects more than 90 species of birds, many of them waders that feed in the intertidal zone on the lee side.

Above left: Farewell Spit is constantly renewed as waves dump sand and wind forms it into dunes before sweeping it out to sea again.

Above right: A sea kayaker prepares to launch from Onetahuti into the clear waters of Tonga Island Marine Reserve. This stretch of coast can be reached only on foot via the Abel Tasman Coast Track, or by water.

Left: Rarangi, on Cloudy Bay in the shadow of snow-capped Tapuae-o-Uenuku, is a popular day trip from Blenheim.

Above: A memorial on Wakefield Quay on the Nelson waterfront honours the strength and tenacity of New Zealand's seafarers.

Right: The historic waterfront along Nelson's Wakefield Quay, refurbished and reinvented after a new commercial port was established nearby.

Left: Picton Harbour, the South Island terminal of the inter-island ferries, is an arm of Queen Charlotte Sound.

Below: When in 1863 the brigantine *Delaware* was wrecked in the bay north of Nelson that now bears its name, local woman Huria Matenga became a national hero by diving into the surf to help rescue the crew.

Right: The many bays and bush-clad islands of the Marlborough Sounds were formed when rising sea levels after the last ice age inundated the coastal valleys.

KAIKOURA

In a dramatic setting locked between the mountains and the Pacific Ocean, cut off by hills to both north and south, isolated Kaikoura's existence is inextricably tied to the sea. The name of the pre-European stronghold gives away its appeal – the Maori name Kaikoura means 'eat crayfish' – and when early European farmers failed they too turned to the water and created a fishing port. Today a booming tourism industry is built around the sea. Swimming with dolphins and sea lions, reef diving and sea kayaking all vie for a place on the active visitor's to-do list, while the less energetic can take boat trips to view sea birds and marine mammals including sperm whales.

Left: The Seaward Kaikoura Mountains rise close behind the windswept town of Kaikoura.

Above left: New Zealand fur seals are a common sight again on the rocky shoreline around Kaikoura. First a source of food for Maori settlers, then a source of wealth for European sealers, they were headed for extinction in the mid 19th century.

Above right: Steam locomotive Ab663 rattles along the spectacular Coastal Pacific route between Christchurch and Blenheim.

Right: Hundreds of dusky dolphins gather off the Kaikoura coast and perform in their usual exuberant style.

WEST COAST

Captain James Cook called the starkly beautiful west coast of the South Island 'wild, craggy and desolate' and the description still holds today. It has never been an easy place to live, what with the difficult access, high rainfall, regular flooding and sandflies, but the abundant natural resources have pulled in a hardy breed of settlers. First came Maori to collect pounamu (greenstone) to make tools, then the hard-living Europeans came in search of gold, followed by coal, timber and fish. As these resources have diminished, the region's last important natural resource, its stunning coastal landscape of lonely beaches and lush forests, has come into focus with the rise of tourism.

Left: The Heaphy Track is one of New Zealand's oldest, longest and busiest tramping – and more recently mountainbiking – tracks. It runs along the Buller coastline between nikau palms and red-flowering rata trees.

Right: At high tide seawater explodes through blowholes in the mysterious pinnacles of stratified limestone at Punakaiki known as the Pancake Rocks.

Below: At Koura Beach, near the Heaphy Track, the powerful waves of the Tasman Sea have lifted up and deposited a bank of boulders.

Above: In the mist of early morning, kahikatea trees rise from a flax swamp behind Gillespies Beach. Vast stands of kahikatea – New Zealand's tallest forest tree – dominated the West Coast's wetlands until European settlers claimed the fertile lowland areas for farming.

Right top: Bleak and weather-beaten, remote Gillespies Beach was torn apart in 1865 by prospectors searching for gold in the radioactive black sand. The miners returned several times over the next century but never with as much success as in the first rush.

Right below: Serpentine Creek drains the coastal plains and enters the sea at Kumara. This section of beach is famous as the start line for the annual Coast-to-Coast multisport race, in which athletes bike, run and kayak over the Main Divide to the beach at New Brighton, Christchurch.

Left: The lush lowland rainforest above Murphys Beach thrives in the mild temperatures and high rainfall of Westland. The West Coast has very few natural harbours and small indented bays like this one provide poor shelter for boats.

Above: The Fiordland crested penguin/tawaki nests on the beaches of the south-western coast of the South Island. Crested penguins are unique among penguins in that they lay two eggs at a time but rear only one of the chicks, an evolutionary quirk scientists are yet to explain.

Right: A number of rivers in South Westland are open to sport fishing. In spring trout are easy to find lurking near the river mouths and making a meal of whitebait swimming up their home river.

CANTERBURY

Much of the Canterbury coast is featureless open beach, sandy in north and shingly in the south. Neatly dividing the two is Banks Peninsula, the eroded remnant of two extinct overlapping volcanoes jutting out from the coast near Christchurch. The first settlers of the region were Ngai Tahu, who naturally lived by the sea where food was to be found, followed in 1840 by the French, who planned to set up a French colony in Akaroa and in the 1850s by the English, who set up large sheep runs on the plains behind the coast. Christchurch, the city established by these English settlers, was severely damaged by earthquakes in 2010 and 2011, including its seaside suburbs of Sumner and Redcliffs and its port, Lyttelton.

Left: Volcanic Banks Peninsula, which Captain Cook marked on his map as an island, once rose 2000 metres above sea level but has been worn down to less than half that height.

Above: Timutimu Head, at the mouth of Akaroa Harbour, is exposed to the worst of the southerly weather.

Right: The historic lighthouse that now overlooks Akaroa was moved in 1980 from its original position protecting the harbour entrance.

Left: The magnificent matai, kahikatea, totara and miro forest that cloaked the hills and lowlands of Banks Peninsula when the European settlers arrived was cleared for pasture. These days, however, tourism has replaced farming as the major industry.

Above: The tiny holiday settlement of Akaroa was once a farming and fishing town. It was founded by French settlers in 1840, and their influence is still felt in the street names and historic houses.

Left, clockwise from top: The beach at Napenape, south of Hurunui Mouth, is brutally exposed to the weather, but some of the best salmon fishing in New Zealand is in the surf just off the river mouth; cliffs eroded from layers of volcanic rock rise from the Pacific Ocean at Akaroa Head; fishers come out in force at the mouth of the Waimakariri River from the very first moments of the whitebaiting season.

Above: Like many along this wild coast, the Motunau Beach settlement is a small one, of fishers and retirees with some holiday houses thrown in. There's an abundance of crayfish and blue cod, and fossil hunters have had good finds in the siltstone cliffs.

OTAGO & THE CATLINS

The coast of the south-eastern part of the South Island is full of variety and abundant with wildlife. North of Dunedin rocky headlands alternate with sandy beaches and bays are few and far between, while Dunedin itself is perched at the edge of the volcanic Otago Peninsula with its penguins, albatrosses and fur seals. To the south those prepared to go off the beaten track into the untouched Catlins will find a string of tiny seaside settlements where the human residents may well be outnumbered by fur seals, penguins nest on the beaches and endangered Hector's dolphins frolic in the bays beyond.

Left: In 1836 a coastal whaling station was set up at Moeraki, which offers rare natural anchorage on the North Otago coast. In a reversal of the usual sequence a Maori settlement followed on the hill behind. Whalers married Maori women, and when whaling stopped being profitable they stayed on and became farmers. Moeraki is now a fishing village and holiday spot.

Above: North of Moeraki the beach is littered with spherical boulders measuring up to 3 metres in diameter. The boulders are geological curiosities, 60-million-year-old concretions formed not by erosion but the gradual build-up of sediments around a small centre.

Left: Taking advantage of its position on a shipping channel not far from the mouth of the narrow Otago Harbour, Port Chalmers is the container port for Dunedin. It also serves as the main South Island port for cruise ships.

Above: A memorial at Port Chalmers commemorates Robert Falcon Scott's ill-fated expedition to Antarctica. In 1910 Port Chalmers was the last port-of-call for his ship, the *Terra Nova*.

Right: Papanui Inlet, on the deeply indented seaward side of the hilly peninsula that protects Otago Harbour, is surrounded by farmland.

Far left: A pou, or carved pole, on the beach at Waikouaiti marks a traditional Ngai Tahu fishing area. Maori had settled the area for many generations before Otago's first Europeans arrived in the 1830s.

Above: Otago Harbour is a drowned remnant of a massive extinct volcano, and Otago Peninsula its eastern flank. The volcano last erupted 10 million years ago.

Below: At high tide Catlins Lake – in reality a saltwater estuary – is rich with trout but at low tide the water all but disappears.

Left: The remarkable Cathedral Caves complex of sea caves soaring up to 30 metres high is accessible only from Waipati beach and only at low tide.

Above: The rock terrace at Curio Bay, complete with petrified tree stumps, is the fossilised floor of a forest dating back 180 million years to the time of the dinosaurs. It is one of the best examples of a Jurassic fossilised forest in the world. The bank behind the bay is a nesting site for yellow-eyed penguins/hoiho, an endangered species that is endemic to New Zealand.

Above: New Zealand sea lions/whakahao like sandy beaches like this one near Nugget Point. This male hasn't quite grown up – mature males have a thick mane around the neck and shoulders. With only 12,000 individuals remaining, the New Zealand species is the world's most threatened sea lion.

Right: Bull kelp/rimurapa flourishes in the cool waters off the southern Otago coast. The tough, flexible blades have an air-filled honeycomb core that keeps them afloat and Ngai Tahu traditionally preserved food in rimurapa.

SOUTHLAND

Southland is a region of bold contrasts, from the plains of fertile farmland around Invercargill to the glacier-formed mountains plunging straight into the sea in the rugged wilderness of Fiordland. The people of mainland New Zealand's southernmost region have always relied on the sea, whether it be early Maori finding food, Europeans (and Maori) hunting seals and whales or farmers sending meat overseas. Commercial fishing, very much part of the Southland identity, began as early as the 1850s and Foveaux Strait oysters were first dredged in the 1860s. Because of the isolation of the communities, each coastal village had its own fishery and some retain a fishing fleet to this day.

Left: Dusky Sound, viewed from Resolution Island, which became a nature reserve in 1891 but sadly the flightless birds relocated there were killed by stoats.

Above: Mitre Peak in Milford Sound/Piopiotahi was carved into its iconic pyramid shape by the glacier that gouged the fiord tens of thousands of years ago.

Right: Doubtful Sound/Patea is less visited than Milford Sound/Piopiotahi, and the more beautiful for it.

Left: Whitebaiters use nets to catch juvenile fish at the mouth of the Mataura River.

Right: New Zealand's southernmost mainland lighthouse looks over the stormy waters of Foveaux Strait from Waipapa Point.

Below left: A signpost at Stirling Point marks the end of State Highway 1's long journey from Cape Reinga.

Below right: The cove aptly named Cosy Nook has several holiday homes and a small fishing fleet.

STEWART ISLAND/RAKIURA

Stewart Island/Rakiura has a maritime past so long it extends into myth: its original name, Te Punga o te Waka a Maui, is a nod to the legend that it was the anchor of demigod Maui's fishing canoe. Maori inhabited New Zealand's third island, catching fish, gathering shellfish and harvesting muttonbirds long before whalers and sealers built shore stations in the early 1800s, and commercial fishing was a mainstay of the economy for over a century. In recent years an influx of visitors unable to resist the island's peaceful sandy beaches and vast wild areas have seen tourism take over as the major industry.

Left: The Rakiura Track, one of New Zealand's Great Walks, winds past and over unspoiled white-sand beaches.

Above: Fishing boats moor in Halfmoon Bay. Fishing for blue cod, crayfish and paua, often by Maori companies, has been a major earner for many decades.

Right: Damaged propellors on the beach at Whalers Base marks the site of the Norwegian Ross Sea Whaling Company's shipyard, last used in 1933.

Above top: Behind the 15-kilometre sweep of Mason Bay beach is one of New Zealand's largest remaining unmodified dune systems, home to a number of threatened native plants and animals.

Above bottom: The anchor chain sculpture at the Lee Bay entrance to Rakiura National Park refers to the island's legendary origins as the anchor stone of Maui's canoe.

Right: Maori birders used the Ernest Islands at the south end of Mason Bay as a final camp before heading out in double-hulled canoes to harvest muttonbirds/titi on the offshore islands.

Also available from Rob Suisted and New Holland

978 1 86966 333 9

978 1 86966 415 2

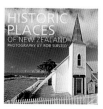
978 1 86966 416 9

978 1 86966 376 6

978 1 86966 332 2

978 1 86966 377 3

978 1 86966 400 8

First published in 2015 by New Holland Publishers (NZ) Ltd
Auckland • Sydney • London

www.newhollandpublishers.co.nz

5/39 Woodside Avenue, Northcote, Auckland 0627, New Zealand
1/66 Gibbes Street, Chatswood, NSW 2067, Australia
The Chandlery, Unit 9, 50 Westminster Bridge Road, London SE1 7QY, United Kingdom

Copyright © 2015 in photography: Rob Suisted,
    www.naturespic.com
Copyright © 2015 New Holland Publishers (NZ) Ltd

ISBN: 978 1 86966 435 0

Managing Director: Fiona Schultz
Publisher: Christine Thomson
Writer: Alison Dench
Design: Thomas Casey
Production Director: Olga Dementiev
Printer: Toppan Leefung Printing Ltd

Front cover: Cape Maria van Diemen and the Motuopao Island Nature Reserve, Cape Reinga.
Title page: Cook Strait waves crash onto the North Island's south coast.
Contents page: Evening at Sandfly Bay, Otago.
Back cover, from top to bottom: Whangamata Harbour; the silhouette of Mount Taranaki (Mount Egmont); the Heaphy Track, along the Buller coastline.

A catalogue record for this book is available from the National Library of New Zealand.

10 9 8 7 6 5 4 3 2 1

All rights reserved. No part of this publication may be reproduced, stored in a retrieval system, or transmitted in any form or by any means, electronic, mechanical, photocopying, recording or otherwise, without the prior permission of the publishers and copyright holders.

While every care has been taken to ensure the information contained in this book is as accurate as possible, the authors and publishers can accept no responsibility for any loss, injury or inconvenience sustained by any person using the advice contained herein.